Heinemann Investigations in Biology

General Editor: S. M. Evans

Investigations in Woodland Ecology

Investigations in Woodland Ecology

C. T. Prime, MA, PhD, FLS, FIBiol.

Former Senior Science Master, Whitgift School

**Heinemann Educational Books Ltd
London**

Heinemann Educational Books Ltd

LONDON   MELBOURNE   TORONTO   AUCKLAND   EDINBURGH
JOHANNESBURG   SINGAPORE   HONG KONG
IBADAN   NAIROBI   NEW DELHI

ISBN 0 435 60283 7

Published by Heinemann Educational Books Ltd
48 Charles Street, London W1X 8AH
Printed in Great Britain by Richard Clay (The Chaucer Press), Ltd
Bungay, Suffolk

To the Biology VIths of
Whitgift School
from whose work this book is
largely written.

'Most of the students have begun the study of Botany in the winter and have acquired the habit of sitting at a bench and examining and drawing fragments of plants. They do not know how to contemplate an entire plant, and consider the union of petals and the number of stamens of greater moment than whether they are dealing with an annual herb or forest tree. It is to the trees they turn for the training their eyes need.'

Humphrey Gilbert-Carter
Director of the Cambridge University Botanic Garden
1921–1950

Preface

This book contains outlines of 31 field-work exercises in woodland ecology which can be carried out with little apparatus. Since the number of possible exercises in plant ecology is very great and comprehensive treatment impossible in a small book, those chosen for inclusion are mainly concerned with the plants of woodlands. However, most of the exercises described can easily be adapted to the study of plants in other habitats. Some of the investigations are straightforward, short and easy, while others are more difficult and require rather more patience and some skill in growing plants (e.g. Investigation 24). None is long term, except perhaps Investigation 22, and this imposes severe limitations on the value of any results obtained, since to establish valid conclusions in ecological work requires more labour than in most other branches of biological science. Any results should be subjected to vigorous criticism in class discussion, otherwise many false or invalid conclusions may be drawn, for it is all too easy to simplify the complex relationships of organisms in nature. In some cases the results should be tested by statistical treatment. No account of statistical methods is given in this book, for it was thought better to describe more exercises rather than to give inadequate mention of statistical mathematics fully treated in many other and readily available books.

There is little included of what might be called the more physiological side of ecology, because the student's time hardly allows him or her to carry out the necessary experiments to gain any data worthy of discussion or argument. One determination, say, of exchangeable calcium in a soil takes a long time, and by itself conveys very little; a large number are required to enable any provisional conclusions to be drawn.

I cannot overstress the value of observing plants in the wild, in the garden, and in laboratories whenever possible, for only in this way can the necessary background knowledge be acquired which saves time and gives perception to biological studies. A student can plan and set up a beautiful series of experiments on the germination of ash, all to be complete by the time of his examination, and find no results if he is unaware that the majority of the seeds only germinate after two winters. Another can make a lot of mistakes by assuming each stem of dog's mercury to be a separate plant, unless he digs one up and finds the matted rhizomes. The sorrow of teachers who find their students have done a lot of work to no purpose is all too frequent.

But for all the difficulties, there is something especially pleasing about much of field work. One can say to the student that no one has done the same thing as he is going to do, for the material with which he is working and the conditions surrounding are so variable that in some ways his work is bound to be different. In this lies both its difficulties and its special fascination.

There remains to say that the faults and errors of this book must be all mine own. I can only hope that they are not too many, and express my thanks to Drs Cunnell and Evans who have read the typescript and made many helpful suggestions. I am grateful to The Clarendon Press, Oxford, for permission to use a quotation from *Our Catkin-Bearing Plants* by H. Gilbert-Carter, 2nd Edition, 1932. Lastly my thanks are due to the patience of Messrs Heinemann who have so readily turned my indifferent typescript into a printed text.

Farleigh, Warlingham, Surrey
1970

Contents

List of Plates

Plate 1 *facing page 2*
Quercus petraea growing on the Blackheath Beds. Notice the absence of shrub layer.

Plate 2 *facing page 3*
Quercus robur growing on the Thanet Sands. Notice the presence of a well-developed shrub layer.

Plate 3 *facing page 18*
A woodland community showing large 'pioneer' beech trees. When one such tree dies, many saplings spring up, forming a 'reproduction circle' (see Plate 4).

Plate 4 *between pages 18 and 19*
A reproduction circle.

Plate 5 *between pages 18 and 19*
A woodland showing cleared coppice in the foreground and well-grown coppice in the background.

Plate 6 *between pages 18 and 19*
(*top*) Male and (*bottom*) female plants of dog's mercury (*Mercurialis perennis*).

Plate 7 *facing page 19*
(*top*) Common mayweed (*Tripleurospermum maritimum*) and (*bottom*) eyebright (*Euphrasia nemorosa*)

Plate 8 *facing page 34*
(*top*) Lesser celandine (*Ranunculus ficaria*) and (*bottom*) wood sorrel (*Oxalis acetosella*)

Plate 9 *facing page 35*
(*left*) A birch tree infected by *Polyporus betulinus* and (*right*) the fruiting body of *P. betulinus*.

1

The plant community

Naturalists of an earlier time were often content to name the plants that they found and to compile lists of species from different habitats. Plant science was not then sufficiently developed for many further investigations to be carried out. Today the scientific observer is no longer content with a record, for he wishes to inquire how the different species are able to live where they do and how they interact and affect each other. Moreover, in contrast to earlier students he has new and powerful scientific techniques with which to tackle the problems that arise. There are, for example, refined methods of soil analysis, accurate ways of measuring light intensities and statistical techniques, to mention only a few that were unknown to the early naturalist. None the less, the early naturalists knew their plants well. Only someone really familiar with the roots of plants could have coined the name devil's bit for *Succisa pratensis* or given the name solomon's seal to *Polygonatum*. The name broomrape (*Orobanche* spp.) betrays an understanding of the parasitic habit and the frequency of the word 'bane' in plant names shows how well the poisonous character of some plants was realised.

Such knowledge gained by observation is greatly to be valued and no student should neglect or despise it. Further, when J. W. White writes in *The Bristol Flora* (1912) that 'the influence of temperature and weather conditions on spring flowering trees is well shown by this species' (the whitebeam, *Sorbus aria*) he is giving ecological information of interest which may well be very relevant to the establishment and survival of this species. This kind of information comes from close study of plants over a long period and there is no substitute for it. It points the way to further scientific investigation, and it picks out the problems most likely to lead to solutions of value to progress in ecology.

Modern studies are therefore deeper and more intensive than those of the past, and it is only possible to make limited studies of this kind. Broadly speaking, it is possible to study the inter-relationships of plants growing in communities (synecology) or to study intensively the ecology of a single species (autecology). These are really two extreme ways of study and one grades into the other. But whichever procedure is adopted it is necessary to choose places or species for study. Likely places, such as woodlands, downlands, marshlands and rivers should be visited with an eye open for areas that are not too much interfered with by man and for areas varied in aspect and rich in plant and animal life. How-ever, it must not be forgotten that some interesting studies can be made of seemingly dull habitats like playing fields, lawns, parks, waste ground and old walls, and there really are very few places where opportunities do not exist.

Natural and semi-natural vegetation is made up of recognisable groups called plant communities. Thus, there are woodland communities, grassland communities and so forth; any chosen area may consist of more than one community when each may be studied in turn. Having made the choice of area it is as well to describe and map it, at least roughly, on a large scale; to state the size, the aspect, the slope, the drainage and the kind of soil; to mention the chief plants and animals; and to find out something of its past history and ownership. Following in the steps of the early pioneers in plant study, a list of the plants growing there should be made.

Investigation 1: To compile a list of species for a chosen (woodland) community

This involves search, collection and identification of the plants. There are many aids to naming plants and they need not be described in any detail. Briefly there are books with illustrations which are widely used, there are floras and there are herbaria

Plate 1. *Quercus petraea* growing on the Blackheath Beds. Notice the absence of shrub layer.

Plate 2. *Quercus robur* growing on the Thanet Sands. Notice the presence of a well-developed shrub layer.

whereby the plant in question can be compared with a named dried or pressed specimen. Herbaria are comparatively little used by beginners but they are very valuable aids to the easy recognition of common and sometimes confusing species. For instance, a named collection of the common grasses is most valuable for their ready identification. It is easily made by pressing the plants between newspaper and they need not be mounted provided a label is attached to each. Note that an herbarium should not be a collection of rare species.

Having compiled the list of the plants in the chosen area, much additional information can be added to it. The frequency of occurrence of the species can be judged subjectively using the following symbols:

d   dominant (very abundant)

a   abundant

f   frequent

c   common

l   local (this letter can also be used in front of the first four categories)

o   occasional

r   rare

v.r. very rare

A rather more detailed procedure has been proposed by Braun-Blanquet (1927) and in this two scales are used. The first gives an indication of the number and cover of a species (abundance) and is as follows:

$+$ sparsely or very sparsely present; cover very small

1   plentiful but of small cover value

2   very numerous or covering at least $\frac{1}{20}$ of the area

3   any number of individuals covering $\frac{1}{4}$ to $\frac{1}{2}$ of the area

4   any number of individuals covering $\frac{1}{2}$ to $\frac{3}{4}$ of the area

5   covering more than $\frac{3}{4}$ of the area

The second scale is used to assess the grouping of the species (sociability) and is as follows:

1 growing singly; isolated individuals
2 grouped or tufted
3 small patches or cushions
4 in small colonies, in extensive patches or forming carpets
5 in pure populations

The one or two, sometimes more, species which are most common are called the dominants, and the presence of these is a feature of a plant community. The word dominant is also used to mean the species which exerts most influence in the plant community; often it is the most abundant species, but this is not necessarily so.

The life form should also be given; details of the Raunkiaer classification of life form will be found in Clapham, Tutin and Warburg (1962). The plant should be studied to find out the life form for some plants can perennate in different ways if they are growing, say, in Cornwall rather than Caithness. An f should be added if the species is in flower and space left for additional information. Thus the final form of the list might be something like that in Table 1. Lists of this kind may be compiled for more

TABLE 1

List of species for Littleheath Woods (May 1969).

Name	Frequency	Abundance (B.B. scale)	Sociability	Life form	Notes
Trees					
Common oak (*Quercus robur*)	d	5	1	Ph	Very few seedlings
Silver birch (*Betula pendula*) etc.	f	2	1	Ph	Absence of *Betula pubescens*?
Shrubs					
Hazel (*Corylus avellana*) etc.	a	3	2	Ph	Heavily galled
Herbs					
Bluebell (*Endymnion non-scriptus*) etc.	a	4	4	G	More abundant to the north of the wood.
Herb bene't (*Geum urbanum*) etc.	1	1	1	Hc	Only along the sides of the paths

Ph = phanerophyte   G = geophyte   Hc = hemicryptophyte

than one area, and if this is done for two or three neighbouring oak woodlands the lists will be very similar, and this shows that a plant community shows a certain constancy of composition. Further, a plant community contains a few species that are very common, and a comparatively large number of species that are rare. In an oakwood there may be ten different species of tree but the oaks may account for 90% or more of the whole. Jaccard (1912) defined the

$$\frac{\text{Number of species common to two areas}}{\text{The total number of species}} \times 100$$

as the coefficient of community, and this may be used to assess the degree of similarity between two areas of vegetation.

Investigation 2: To determine the coefficient of community for two woodlands

Make lists of the species for sample areas, e.g. 5-metre squares in the two woodlands chosen, and calculate the coefficient in each case. Repeat this procedure using sample areas of increasing size. Plot a graph of the coefficient of community *v.* the size of the sample area and interpret your results.

These lists will show that the plant community shows a relation to the habitat, i.e. that different habitats are occupied by different species. It is, however, clear that lists of this kind give an inadequate idea of the vegetation as a whole, for they give only an estimate of the relative abundance and little idea of the detailed distribution of the species of which it is composed. Methods less dependent on subjective assessment have therefore been thought out.

Investigation 3: To sample a plant community by quadrat and transect

To attempt to assess distribution, abundance and cover many sampling methods have been devised. One of the most widely used is the quadrat. Fundamentally this consists of a small area which is carefully mapped in detail, each plant or shoot being represented by a letter or symbol. Thus an area of 1 square metre can be marked out on the ground with the aid of four pegs and a length of cord, and the individual plants recorded on squared paper. Several questions arise in the use of this method; they concern the position, the size and the number of quadrats to be made in order to gain a reliable picture of the vegetation. The samples can be taken at random and this is often attempted by throwing something over the shoulder and marking the quadrat wherever it falls. Actually such a practice does not necessarily give a truly random distribution but it will serve for all practical purposes. Alternatively, the sampling can be systematic; in which case a grid is marked or paced out and the quadrats carried out

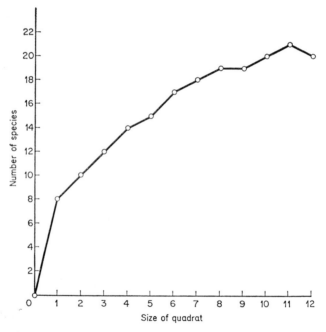

Figure 1. A graph showing the number of species per quadrat plotted against quadrat size.

at the points of intersection. It is instructive to compare the results obtained by different methods of sampling, and for a full discussion of this the reader is referred to P. Greig-Smith (1964).

The problem of size is a little more difficult, for plainly if one is interested in sampling the tree population, the size must be larger than if one is sampling the field or ground layer of a wood-land. This matter can be decided by making quadrats of gradually increasing size and recording the number of species in each. A graph is then plotted of the number of species against the size of the quadrat and a result is shown in Figure 1. It will be seen from this graph that at the point where the curve flattens the majority of the species will be included in a sample of this size. This is known as the minimal area; if we halve the area we lose five of the species, but if we double it we hardly gain any. The species area curve therefore indicates the size of the quadrat suitable for making species lists; it does not necessarily give guidance as to the suitable size of the quadrat for estimating abundance, cover or sociability which are quite different things

There are many variants of the simple quadrat and one commonly used for the field layer of a woodland or a grass sward is a point frame. This is a wooden frame of crossed strips; each is divided into ten, and at each point a small hole is bored through which passes a stiff point (e.g. a knitting needle). There are therefore 100 points in all. The frame is placed on the ground and any individual touching the needle point is recorded; the results can be expressed as percentages. Cover can be estimated by the eye and owing to the fact that individuals can overlap, the total cover may very well add up to over 100%. (See Table 2.)

Study of the results brings out another feature of the plant community, namely that species tend to occur together and to show an inter-relationship. Bluebell (*Endymnion non-scriptus*) often occurs in association with bracken (*Pteridium aquilinum*) for the fronds of the latter come through the ground as the bluebell finishes flowering and the bulbs of the bluebell exploit a region of the soil below the level of the rhizome of the bracken. This is also an example of stratification, namely that in a plant community the vegetation is arranged in tiers. Nowhere is this better shown than in a woodland which may show a tree layer, a shrub layer, a field layer and a ground layer. Sample methods which record this in detail are called transects; in this case a line of string is drawn across the area and the individuals touching it are recorded to scale by means of suitable symbols. If the individuals are too

TABLE 2

The distribution and coverage of plants in the field layer of a woodland.

Plant	Point frame analysis (percentage)	Cover (est. percentage)
Blackberry (*Rubus fruticosus*)	5	9
Bracken (*Pteridium aquilinum*)	4	5
Wood sorrel (*Oxalis acetosella*)	10	10
Bluebell (*Endymnion non-scriptus*)	31	40
Soft grass (*Holcus mollis*)	7	10

numerous to record, then only those present at fixed intervals need be noted. Here, to appreciate the stratification it is desirable to decide on suitable vertical and horizontal scales. (See Figure 2.)

Finally, plant communities show a relationship to time; this can only be studied by observation over a considerable period. Some changes are quite obvious, such as the successional changes on waste ground as it becomes colonised by scrub and woodland. But plant communities which appear permanent and unchanging, like those on chalk downland, will reveal a changing pattern if permanent quadrats are put down and carefully recorded from year to year. These studies serve to emphasise the dynamic nature of vegetation.

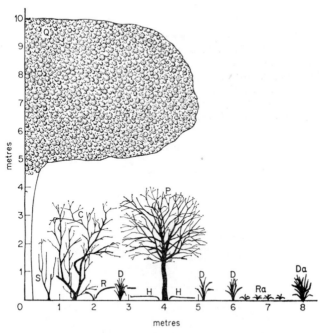

Figure 2. A transect through woodland, showing stratification. (Only the major species are indicated.)

Observation and study of the kind indicated in this chapter will bring to light many unsolved problems worthy of further study. The notes added against the species listed in Table 1 are indicative of the kind of problems that arise. Which of these should or can be taken further? The student must decide this for himself. He must grasp the nature of the problems and decide how far they might be solved by experimental attack and see if the necessary work can be done with the resources he has and within the limits of the time available. Here judgement and insight are of great importance, and they are aided by preliminary and careful study of the areas under investigation. What is quite certain is that once he tackles any work of this kind, the student will find no lack of problems to solve.

References

Braun-Blanquet, J. (1927). *Pflanzensoziologie*. Springer, Wien.
Clapham, A. R., Tutin, T. G. and Warburg, E. F. (1962). *Flora of the British Isles*. C.U.P.
Greig-Smith, P. (1964). *Quantitative Plant Ecology*. Butterworth.
Jaccard, P. (1912). *New Phytologist*, 11, 37–50.
Kershaw, K. A. (1964). *Quantitative and Dynamic Ecology*. Arnold.
White, J. W. (1912). *Flora of Bristol*. Wright, Bristol.

2

Tree distribution in woodland

Comparatively few woodlands show a uniform tree composition and distribution. This is only to be found in a plantation where the same species has been planted in rows over the whole area, but even here the actual growth and size of the trees may come to reflect differences in the aspect and slope of the ground, in the soil composition and in the features of the environment. All these differences may well be worthy of study. Usually our woodlands are far more varied than plantations both in their composition and structure; many are long established and hence they reflect changes of management as well as those due to natural factors like soil, climate and the activity of animals.

Many patterns of distribution will be found. For instance, where two types of soil meet there may very well be a more or less sharp alteration in the tree pattern. In the south of England where the chalk is overlain in places by clay-with-flints, the latter often carries an oak woodland whereas the chalk bears a beechwood. There may be further differences in composition within the one kind of woodland. Some beechwoods contain much yew, while in others it is comparatively rare. Scrub developing on a neglected chalk slope may be almost pure hawthorn or it may be quite varied in composition, with whitebeam, wayfaring tree and other shrubs playing a part.

In other woodlands there may be a closer interaction between the dominant trees as when two related species meet and hybridise. This happens with the two species of oak, the common or pedunculate oak (*Quercus robur*) and the durmast or sessile oak (*Q. petraea*), the two species of birch (*Betula pendula* and *B. pubescens*), the two hawthorns (*Crataegus monogyna* and *C. laerigata*) and the several species of willow (*Salix* spp.). As an illustration of this type of investigation the distribution of oaks in a woodland known as Croham Hurst has been studied and is outlined below. In passing it may be said that the two oaks are not the easiest of trees to study in this way because they are very variable species in themselves.

Croham Hurst is a woodland containing several species of tree, but the dominants are the two oaks mentioned above. The woodland spreads over two geological formations, one a ridge of Blackheath pebble beds which in turn rests upon and is surrounded by the second, a bed of Thanet sand. The boundary between the two is marked by a sharp change of slope. *Quercus robur* occurs mainly on the Thanet sand at the base, while *Quercus petraea* is found on the central ridge of Blackheath pebbles, see Plates 1 and 2. In the intervening region hybrids between the two species occur.

Investigation 4: To study the distribution of the two species of oak and their hybrids

The two species of oak differ in about twelve different ways which are listed in Table 3. Some of these are illustrated in the drawings of the two shoots (Figure 3).

In order to assess the degree of hybridity it is necessary to measure each of the differences where possible. This is easy to do for the quantitative differences like the length of the acorn stalk but clearly impossible for others such as the degree of hairiness, These latter differences may be assessed by grading on a point scale. Thus, for example:

> deep lobing of the leaf may be assessed as   1
> intermediate as   2
> shallow as   3

and strong auricle of the leaf base as   1

> intermediate (near 1) as   2
> intermediate (near 4) as   3
> no auricle as   4

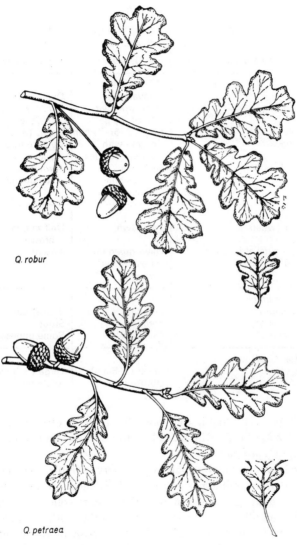

Q. robur

Q. petraea

Figure 3.

TABLE 3

The differences between the two species of oak.

Character	Q. robur	Q. petraea
Leaves		
Shape	Obovate	Ovate
Lobing	Deep, irregular 3–5(–6) pairs	Shallow, regular, 5–6(–8) pairs
Petiole	Short, 2–3(–7) mm	Long, 13–25 mm
Leaf base	Cordate with strong auricles	Cordate to cuneate with weak auricles
Abaxial surface	Glabrous, occasional simple hairs	Always some stellate hairs
Acorns		
Colour (when ripe)	Pale fawn	Uniform dark brown
Longitudinal stripes (when ripe)	Olive-green on fresh mature acorns	Absent
Fruiting peduncle		
Total length	2–9 cm	0–3(4) cm
Thickness	Slender	Stout
Pubescence	Glabrous	Some pubescence
Buds (terminal)		
Size	Small (<5mm)	Large (>5 mm)
Apex	Obtuse tip	Acute tip

It will be seen that if the totals for these grades are added, *Quercus robur* forms will have low totals while high totals will indicate *Q. petraea* forms.

The leaves on any one oak tree are very variable, as can easily be seen by examining a few trees. It is therefore desirable to adopt a standard method of collection, e.g. to collect leaves, say, from the middle of a fully grown shoot on the south side of a tree. Leaves from shoots springing from the base of the trunk, or lammas (summer) shoots are particularly variable and should not be collected. If necessary, a long-arm pruner should be used to gather the sample, and as many of the characters mentioned in Table 3 should be measured or assessed for at least ten leaves or acorns from each tree. To make a complete assessment it will be necessary to collect leaves in summer, acorns in autumn and buds in the winter, but time may not allow this.

Having made a decision as to how many characters are to be recorded, make a transect 5 metres wide across the area, recording the data for each tree met with in the transect; if possible, repeat this so that the area in question is adequately sampled. Incidentally, the question of what constitutes adequate sampling in an investigation of this kind can be made the subject of further, or better still, preliminary investigation.

The sum of the grades for each sample must be added together to give a total grade for each tree; these totals are grouped, and a histogram plotted of grade total erected against the frequency of the trees having a particular grade total. Further separate histograms may be drawn for the trees present in the various areas investigated; in the particular case for the oaks on the Blackheath pebbles and for those on the Thanet sand.

The quantitative data may be treated differently. A scatter diagram can be plotted for petiole length as a percentage of total leaf length against the position in the transects. Alternatively the data can be handled by statistical methods.

The result in question shows evidence of a change from *Quercus robur* at one extreme to *Quercus petraea* at the other. It is, however, clear that a large number fall into the intermediate grade and few are true *Querus robur* and few true *Quercus petraea*. The interpretation of such a result is a matter for discussion, but when two formerly isolated populations re-establish contact before too marked a genetic divergence has occurred, crossing between them leads not to reproductive isolation but to introgression. In each case variable populations exhibiting characteristics of each will be found throughout a wide area of overlap. Such introgressive hybridisation would explain the results shown in Figure 4, but for such an interpretation confirmatory evidence is desirable and historical records should be looked into with this end in view.

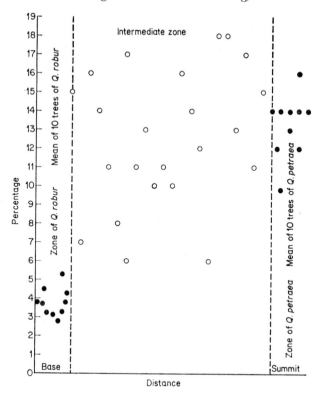

Figure 4. Simplified scatter diagram of the length of the petiole as a percentage of leaf length plotted against the position of the tree on the slope. The zone at the top is almost pure *Q. petraea*, the zone at the base contains some hybrids (not shown in the diagram) while the intermediate zone contains *Q. petraea* and some hybrids.

Detailed historical research is a discipline of its own but useful information can often be gained by consulting old maps at the nearest public library. As far as Croham Hurst is concerned, the name hurst means a woodland, and the name is a very old one. On maps dated 1720, 1749, 1761 and 1820, the area is shown as woodland, and it seems fairly certain that in this small area we have a small piece of relic woodland that has never been clear-felled. Thus the present distribution has come into being over a long period of time.

This investigation is but one example of a particular kind; other woodland trees or herbaceous species can provide similar studies, with quite different results. For example, at Croham Hurst the distribution of the two birch trees (*Betula pendula* and *B. pubescens*) does not show any relationship to the underlying geology. Among herbs, the campions (either the red and white, *Silene alba* and *S. dioica*) or the bladder campions (*Silene cucubalus* and *S. maritima*) suggest themselves as suitable plants for this kind of study.

Most woodland populations, however permanent they seem, are in a state of change for their relationships to the environment are dynamic and not static. Even if woodland appears the same over a long period, there are changes going on all the time for some trees are dying and being replaced by others. The death and fall of a large tree leaves a big gap in the tree canopy. This gap is rapidly invaded by many plants and great competition ensues. Gradually the number of individuals decreases until perhaps only one replaces the original. The process, in part a repetition in miniature of the original succession by which the woodland became established, is usually spoken of as a reproduction circle, see Plates 3 and 4. This process can be studied by recording the ages and number of each type of tree seedling in such a circle.

Investigation 5: To study the number and ages of horse chestnut (*Aesculus hippocastanum*) saplings in a reproduction circle

Record the number of tree seedlings, including saplings, in a reproduction circle, and in each case determine the age by counting the ring scars on the main axis only. This is easy to do for the first few years because the ring scars are obvious, but later they become obscure; then it is permissible to make an estimate based on the lengths of the previous annual increments.

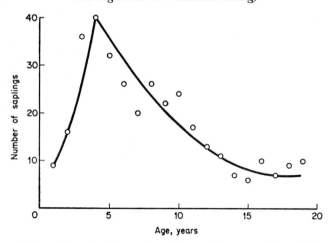

Figure 5. Graph showing the numbers of horse chestnut saplings of different ages in a reproduction circle.

Plot a graph of the number of each species of tree against the age.

Figure 5 shows the number of horse chestnut seedlings of different ages in a reproduction circle in a mixed plantation. The reproduction circle was brought about in the first place by the fall of a large horse chestnut tree. How do you account for the small number of young seedlings 1–2 years old? Can you suggest reasons for the decline in numbers after 4–5 years?

Investigation 6: To study the annual growth in length of horse chestnut saplings

Measure the annual increments of ten of the larger horse chestnut saplings in a reproduction circle. What can you deduce from your measurements? At what age is competition most intense in this example?

In any woodland if the dominant species is constant, one might expect the younger or older saplings to reflect the pattern of the adult trees and conversely. On an oakwood of *Quercus robur*, one

Plate 3. A woodland community showing large 'pioneer' beech trees. When one such tree dies, many saplings spring up, forming a 'reproduction circle' (see Plate 4).

Plate 4. A reproduction circle.

Plate 5. A woodland showing cleared coppice in the foreground and well-grown coppice in the background.

Plate 6. (*top*) Male and (*bottom*) female plants of dog's mercury (*Mercurialis perennis*).

Plate 7. (*top*) Common mayweed (*Tripleurospermum maritimum*) and (*bottom*) eyebright (*Euphrasia nemorosa*).

might expect to see a similar proportion in the young trees. Conversely, in a neglected plantation of mixed species, change might well be expected, since it is known that many introduced and frequently planted trees do not maintain themselves in this country.

Investigation 7: To compare the composition of the adult tree population of a plantation with the composition of the sapling population

Devise methods to study and record the distribution of the adult trees in a plantation. Repeat these observations, studying the younger trees or saplings which have obviously grown since the original establishment of the plantation, but neglect small seedlings. From your results forecast the possible changes in the composition of the plantation as it progresses towards woodland.

References
Carlisle. A. and Brown, A. H. F. (1965). The assessment of the taxonomic status of mixed oak (*Quercus* spp.) populations. *Watsonia*, 6, 120–127

3

The shrub and field layers of woodlands

Woodlands provide the best example of stratification in vegetation that can be seen in Great Britain. In many, particularly oak woodlands, there are at least four layers, namely the trees, the shrubs below, the herbs of the field layer and the mosses and small plants of the ground layer. In oak woodlands the shrub layer is known as coppice, because it was formerly cut for a great variety of purposes like making charcoal for use in iron smelting or for use as hurdles, fencing and poles of several kinds. Coppice in *Quercus robur* woodland consists mainly of hazel and other shrubs which sprout readily from the base when cut; this was usually done at intervals of about 15 years. Coppicing is still carried out in many woodlands, and the shrub clearance is followed by a great development of the field layer which gradually declines as the shrubs grow up to their full size once more, see Plate 5.

Investigation 8: To study the results of coppicing on the field layer of a woodland

Devise methods to compare the frequency and vigour (e.g. height and size of inflorescence) of selected species in (*a*) woodland showing well-grown coppice and (*b*) woodland in the second year after the coppice has been cut. Suitable species for this study might be the marsh thistle (*Cirsium palustre*), the foxglove (*Digitalis purpurea*), the mullein (*Verbascum thapsus*) and the rosebay (*Chamenerion angustifolium*), as well as the more frequently occurring members of the field layers.

The field and ground layers of woodlands are very varied in

composition, ranging from the sparse carpet of a young beech-wood to the numerous patterns made up of large numbers of individuals seen in oak woodlands. In the oak woodlands dominated by *Quercus robur*, Tansley (1939) has described several subordinate communities or societies of plants. Perhaps the first to mention is that dominated by the bluebell (*Endymnion nonscriptus*) which forms sheets of glorious blue flowers mainly on the lighter soils. Incidentally, the bluebell is an Atlantic species confined to West Europe, so its sheer magnificence is seen at its best in Great Britain. Often the bluebell is associated with the wood anemone (*Anemone nemorosa*) though this plant avoids the driest and wettest soils, and is perhaps most characteristic of the lighter and medium loams.

On the lighter soils in woodland dominated by either species of oak, a society of bracken (*Pteridium aquilinum*), soft grass (*Holcus mollis*) and bluebell is quite frequent. In this society, the three species develop one after the other, bluebell being the first to appear, followed by soft grass and later by the bracken. Also, the underground parts develop at different depths and in this way competition between each is minimised. On the heavier and more basic soils, dog's mercury (*Mercurialis perennis*) is a frequent dominant, but this plant avoids acid soils and those which become waterlogged in winter. Cuckoo-pint (*Arum maculatum*) and lesser celandine (*Ranunculus ficaria*) are species often associated with it.

On the very damp soils, creeping buttercup (*Ranunculus repens*), tufted hairgrass (*Deschampsia caespitosa*) and meadow-sweet (*Filipendula ulmaria*) may be the dominants of societies together with plants like angelica (*Angelica sylvestris*) and the nettle (*Urtica dioica*). In the wettest parts, sedges (*Carex* spp.) and rushes (*Juncus* spp.) may also play an important role in the vegetation. In well-lit parts of oak woodlands, the red campion (*Silene dioica*), the yellow deadnettle (*Lamiastrum galeobdolon*) and the wood spurge (*Euphorbia amygdaloides*) flower freely, particularly in the years immediately following the cutting of the coppice. As shade increases these species flower less frequently and in low light intensity they remain continually in the vegetative condition. The distribution and frequency of occurrence of these societies pose many interesting problems, as do the inter-relationships of the species of which they are composed. Some of the following investigations are concerned with them.

Investigation 9: The distribution of the bluebell in relation to light intensity

Mark out in a suitable woodland area a grid, say, 20 by 10 metres. At each point of intersection place a circular loop of wire 25 cm in diameter and count the number of bluebell shoots within it. At each point of intersection measure the light intensity with a light meter, and at the same time arrange for the full light intensity in the open to be read. The light intensities at the intersections of the grid line can then be expressed as percentages of those outside. Repeat these observations on separate occasions, and finally plot a scatter of the numbers of individuals against the light intensities.

There are certain problems that arise in the procedure outlined above. Should bluebell seedlings be excluded from the counts of individuals for example? Light intensity fluctuates, owing to variation in cloud cover and to the movement of the leaves of the trees above. How may these factors be taken into account? Suppose also that some kind of association between the number of individuals and light is found in this one experiment, does this amount to proof of a causal connection? What further steps should be taken to establish such a causality?

Investigation 10: To study the effect of light intensity on the flowering of the bluebell

Making use of the lay-out of the previous investigation count the number of flowers per inflorescence and the number of seeds per capsule in areas of differing light intensity. This should be done in uncut and recently coppiced areas, and the data used to compare the performance of the species in the different areas. Also, making additional observations if necessary, work out the yield of seed per square decimetre for the areas you have studied.

Repeat this investigation with appropriate modification for the earth nut (*Conopodium majus*), the lesser celandine (*Ranunculus ficaria*) and the wood anemone (*Anemone nemorosa*). Compare also the dates of flowering of these species in uncut and recently coppiced areas.

Investigation 11: To study the distribution of the male and female plants of dog's mercury (*Mercurialis perennis*) in relation to light intensity (Plate 6)

Dog's mercury is a plant with upright stems which is freely produced from rhizomes and which comparatively rarely reproduces from seed. It also bears unisexual flowers and so clones are found in which all the plants are either male or female. Make preliminary investigations and then devise methods to see if the frequency of occurrence of the male and female plants is related to light intensity.

Investigation 12: To study the distribution of dog's mercury in relation to soils of high water content

Dog's mercury is a plant which does not grow well when the soil is waterlogged for long periods in the year. Investigate the distribution of this plant in relation to the soil water by recording the number of stems of the plant per unit area and measuring the soil water content at regular intervals. This can be done by taking simultaneous samples of about 10 g and weighing each in a weighed basin. Dry each oven at or near 100° C, cool in a desiccator and re-weigh. Continue to do so until there is no further loss in weight. The soil must not be heated too strongly or weight losses will occur due to changes in its composition.

In the more waterlogged areas try the effect of raising the level of the soil by 10 cm and planting some dog's mercury on it. Compare its performance with that in a control area.

Investigation 13: To compare the performance of dog's mercury in uncut and coppiced areas

Collect equal numbers (50–100) of shoots of dog's mercury from coppiced and uncoppiced areas and determine their fresh weight; dry them in an oven at or near 100° C, cool in a desiccator and re-weigh.

Repeat this experiment with the shoots of the bluebell and the shoots of cuckoo-pint. Compare also the fresh and dry weights of bulbs and corms respectively collected from coppiced and uncoppiced areas.

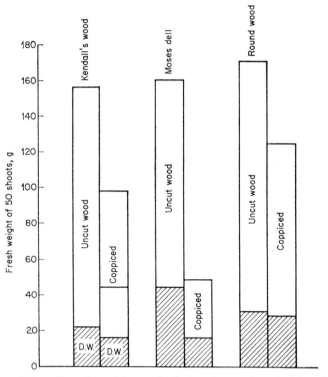

Figure 6. The effect of coppicing on the fresh and dry weight of *Mercurialis*. (Data from Salisbury (1924).)

Investigation 14: To compare the growth of the yellow deadnettle (*Lamiastrum galeobdolon*) in well-lit and dull areas

The yellow deadnettle is a species that reproduces vegetatively by long leafy stolons. Measure the lengths of the stolons produced by individual rootstocks from (*a*) plants in areas where there is no flower and (*b*) from areas where the plant flowers freely. Record also the date and period of development of the stolons in the two areas. Compare the data carefully and devise methods to present any differences observed in a clear and obvious manner.

If the opportunity arises, compare the number and length of the stolons in a dry and wet season. If the plants are in flower, compare also the number of fruits produced in dry and wet seasons.

Investigation 15: To study the growth of woodland plants in relation to root competition

It is tempting to correlate the growth of field layers in woodland with the light intensity all too completely and to forget that other factors like competition between root systems may be almost as important. Mark out areas in a woodland, and in some cut down deeply round the sides with a spade so as to sever the competing tree and other roots. Pieces of slate or glass may be driven down into the soil to prevent the re-entrance of the area by roots. Compare the performance in the experimental plots with those in the controls by measuring (*a*) the fresh weight and (*b*) the dry weight of the plants.

Investigation 16: To study the performance of the cuckoo-pint (*Arum maculatum*) in different woodland areas

Mature cuckoo-pint plants usually have three leaves which start to emerge and unfold in southern England about the first week in February, and which die away by the end of June. The actual period for active photosynthesis is therefore relatively short, since in the early stages temperature may be limiting and at a later stage when the tree canopy is complete, light may be limiting.

Compare the performance of the plants in two different situations (say, north- and south-facing slopes in a woodland) by measuring the heights of the corresponding leaves (first, second and third in order of emergence) at weekly intervals, from the time of their first appearance to the time that they are fully expanded.

These results may show very little if individual variations in leaf size are sufficiently great as to distort the averages. This is therefore the type of investigation where statistical treatment of the data is desirable, see Figure 7.

Early unfolding of the leaves does not confer any biological advantage unless they can carry out effective photosynthesis. If the leaves of the tree canopy also expand earlier, they may clearly so reduce the light intensity as to vitiate any advantage accruing from early unfolding. In interpreting these results, it should be borne in mind that photosynthesis of shade plants is relatively more efficient in moderate light than in high light intensities. (Relate this to the results of Investigation 13.)

Investigation 17: To investigate the association between pairs of species, e.g. anemone and bluebell

Mark out a relatively large quadrat, say, 3 by 3 metres and determine the cover of the two species by point sampling with a frame of 225 points. This is perhaps achieved more easily by using a smaller frame with 15 points 9 times over to cover the whole area. This sample must be repeated several times and the number of occurrences of one plant plotted against those for the other. Further analysis can be carried out by statistical methods. In this investigation, the size of the sample area is important, and it is instructive, if time allows, to repeat the experiment with areas of different size. For a further and more complete discussion of the problems involved, see P. Greig-Smith (1964).

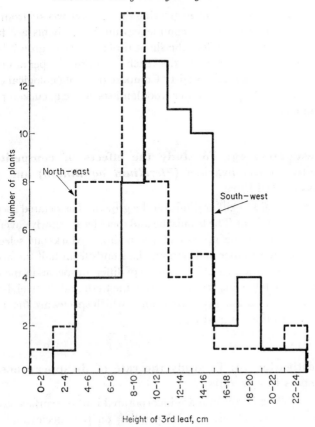

Figure 7. The height of the third leaves of cuckoo-pint on the north-east and south-west slopes of a woodland in mid-March.

Investigation 18: To study the growth of ivy (*Hedera helix*)

Ivy is an evergreen plant which often forms a leafy carpet on the floor of woodlands, and it flourishes in a very low light intensity. The evergreen habit clearly allows the plant to photosynthesise throughout the year if temperatures and other factors are suitable.

Measure the lengths of 20 growing shoots of ivy weekly from the time that elongation first commences until the shoots are fully grown. At what date does the shoot tip first start to grow? How many leaves, on the average, does each shoot produce per annum? How long do the leaves persist? Compare these phenological data with those obtained for other woodland species, e.g. cuckoo-pint, anemone.

Investigation 19: To study the effects of competition between the bracken (*Pteridium aquilinum*) and the associated flora

Bracken is said to kill or diminish the growth of associated species by falling on them. The fronds die and then decay, slowly covering and preventing the growth of competitors. Mark out selected areas in late summer and remove the fronds from half the areas. Compare the number of individuals of different species in the two areas in the following season and also the fresh and dry weights of the total plant mass. Try to determine which species are the more resistant to this competition.

Investigation 20: To study the rate of disappearance of leaves in a woodland

Leaves falling in autumn are incorporated into the surface layers of the soil at varying rates depending on the species and the conditions of the site where they fall. Generally speaking, more organic material accumulates under conifers than under hardwoods, and in pinewoods about 30–40 years of age about half of this is derived from the leaves, and half from cones and twigs. Further, this organic matter is more sharply differentiated from the soil beneath than in deciduous woodlands, and it is more acid in character. It is known as mor. By contrast the organic matter formed under hard woods is less acidic, and is known as mull. Mull tends to contain a higher percentage of nutrient elements such as calcium, magnesium and phosphorus.

Devise sampling methods to study the rate of disappearance (i.e. decay and incorporation into the surface organic matter) in two or more types of woodland during the months following leaf fall in autumn. Which leaves are the first to disappear and which persist the longest? Approximate estimates of the nutrient content may be made by drying leaf samples to constant weight at or near 100° C in an oven and then incinerating them to constant weight in a small furnace.

In general, the increased illumination following upon coppicing brings about a more rapid decomposition of the humus, which results in a decreased water-retaining power in the surface layers of the soil and a rise in acidity. This rise is temporary and does not persist for more than 1–2 years after coppicing. Following this conditions become more favourable to acid intolerant species and more favourable for the formation of nitrate. Many of the species which are frequent in the coppiced areas are plants that require a good supply of nitrates to grow well.

Compare the water-retaining power, the pH (use a capillator) and the nitrate content (use diphenylamine) of surface soils in coppiced and uncut woodland.

For the latter two determinations prepare a soil extract using an extracting solution made us as follows:

Sodium acetate	50 g
Acetic acid (glacial)	15 g
Water	500 cm^3

Add 10 cm^3 of this to about 5 g of soil, stir for one minute; filter and use the filtrate for the tests. For nitrogen transfer one drop of the filtrate to a white tile and add four drops of 0·2% diphenylamine in concentrated sulphuric acid. Compare the resulting blue colours for the soils under test.

Investigation 21: To study variation in the inflorescences of the cuckoo-pint (*Arum maculatum*)

The pollination of the cuckoo-pint is well known and is described in many textbooks. The species is, however, very variable, and any large population of this common woodland plant will show much variation in the colour of the spadix and the degree of spotting of the leaves. The spadix may vary from almost black to quite yellow with many intermediates. Curiously, most species of *Arum* do not show a similar range of variation; in the only other species (*Arum italicum*) to be found in Great Britain it is always yellow. Variation in leaf spotting also occurs, some being entirely without anthocyanin, while others show dark blotches. This variation is an example of what Sir J. Huxley has called morphism (1955). Several localities may be visited and the numbers of plants recorded as under:

Number of plants with purple spadix and spotted leaves
Number of plants with purple spadix and unspotted leaves
Number of plants with intermediately coloured spadix and spotted leaves
Number of plants with intermediately coloured spadix and unspotted leaves
Number of plants with yellow spadix and spotted leaves
Number of plants with yellow spadix and unspotted leaves

The reasons for this variation and its maintenance in nature are quite unknown, but it is valuable to propound possible explanations.

Investigation 22: The biology of the species of *Hypericum*; a suggestion for a more extended study

Hypericum perforatum is the most frequent St John's wort in coppiced woods of *Quercus robur*, while *Hypericum hirsutum* is more frequent in or near beechwood. In the woods of *Quercus petraea* where soils are lighter and more acid, *Hypericum humifusum* and *Hypericum pulchrum* are the most common.

Investigate the distribution of these species in your own neighbourhood to see how far the above generalised statement is true, and devise further experiments and observations to endeavour to account for the recorded distributions.

References

Greig-Smith, P. (1964). *Quantitative Plant Ecology*. Butterworth.

Huxley, J. S. (1955). Morphism and Evolution. *Heredity*, 9, 1–52.

Mukerji, S. K. (1936). Contributions to the autecology of *Mercurialis perennis*. *J. Ecol.*, 24, 38.

Prime, C. T. (1960). *Lords and Ladies*. Collins.

Rutter, A. J. and Blackman, G. E. (1946). The Light Factor and the distribution of the Bluebell in Woodland Communities. *Ann. Bot.*, 10, 361.

Salisbury, E. J. (1924). The effects of coppicing as illustrated by the Woods of Herts. *Trans. Herts N.H.S.*, 18, 1–21

Salisbury, E. J. (1929). The biological equipment of species in relation to competition. *J. Ecol.*, 17, 197

Salisbury, E. J. (1942). *The Reproductive capacity of Plants*. Bell.

Tansley, A. G. (1939). *The British Isles and their Vegetation*. C.U.P.

4

Some woodland predators and parasites

Investigation 23: The infections of acorns by nut weevils (*Curculio* (*Balaninus*), *glandium* and *venosus*)

The nut weevils all possess a very long thin snout by which they pierce a hole in the young acorn or hazel nut. They burrow into the nut and lay an egg in the soft tissue within. Egg-laying takes place in the early summer when the outer tissues of the acorn are still soft enough to allow the proboscis to penetrate them. The larvae mature in the late summer and autumn when the infected acorns tend to fall early. The larvae then leave the nut and pupate in the soil.

Collect a large number of acorns from under either *Q. robur* or *petraea* and separate off those showing a hole where the insect larva has left the acorn for pupation. Usually the infected acorn is slightly discoloured. Calculate the percentage infection by these insects. Investigate the cause of the infection by dissecting some of the acorns. Place some of the infected acorns in sterilised soil in a petri dish and allow the larvae to pupate. Hence try to study the remainder of the life cycle till the imago emerges in the spring. Count out 100 uninfected acorns and germinate them in a large dish under the most favourable conditions and determine the percentage that germinate. Repeat the experiment with infected acorns. Study the morphology of the acorn by dissection to see how it is damaged so that germination is prevented.

Investigation 24: To study the biology of the semiparasitic cow wheats (*Melampyrum* spp.) and eyebrights (*Eurphasia* spp.)

Of the four species of *Melampyrum* to be found in Great Britain one *M. pratense* (common cow wheat) is frequent in woods and on heaths throughout the country, preferring on the whole the drier and lighter soils. It is an annual and it is parasitic on the roots of

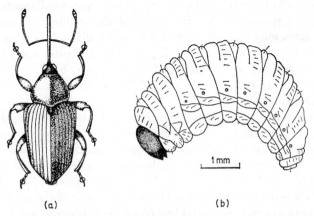

(a)   (b)

Figure 8. (*a*) *Curculio* sp. (after Escherich) and (*b*) larva of *Curculio* sp. (after Scherf).

shrubs. The eyebrights (see Plate 7) are common annuals and are to be found in a variety of situations. The species are difficult to determine, but as a group they are easily recognisable. *E. nemorosa* is a common species of woods, downs and pastures and it may well be more readily studied than *Melampyrum pratense*. The seeds of *Euphrasia* are small and numerous but *M. pratense* produces comparatively few seeds.

Record the occurrences of colonies of the common cow wheat and note particularly the type of soils where they are to be found. Record also the associated trees and shrubs, and also the neighbouring herbs. If the size of the colony allows it, dig up a small block of earth with the plant in it and remove it to the laboratory.

Wash very carefully under the tap to find the attachments between the parasite and the roots of the host. This is difficult to do and requires much care. Does the parasite confine its attention to one species only?

Determine the number of seeds produced by individual plants and consider this in relation to the frequently made statement that parasites produce a very large number of propagules. Germinate some of the seeds in sterilised soil. Seeds germinate fairly well and produce two or three pairs of very small leaves but unless a parasitic union is established the seedlings die away. Attempts may be made to establish such a union by sowing the seeds in soils containing the living roots of possible hosts (e.g. *Corylus avellana, Betula pubescens, Calluna vulgaris*) but this is difficult to do. In only a small fraction of the seedlings is a successful union achieved.

The seeds of *Euphrasia* may be germinated in spring after exposure to winter cold; seed loses its viability if kept for months in too dry conditions. *Euphrasia nemorosa* can be grown with the plantain (*Plantago lanceolata*) as host; it is desirable to raise the plantains first as small plants and sow the *Euphrasia* around it. Although the roots of the two plants forms a very dense mass, it is not easy to make out individual connections between the two.

Investigation 25: To study the infection of the lesser celandine (*Ranunculus ficaria*) by rust disease

The leaves of the lesser celandine (see Plate 8) are frequented by a rust fungus *Uromyces ficariae*. The pustules occur on the underside of the leaves and under the microscope will be seen to consist of stalked, one-celled teleutospores, each with an apical germ pore. Another species of *Uromyces*, *U. poae* produces uredo and teleutospores on *Poa* spp. and aecidia on some species of *Ranunculus* including *R. ficaria*, but this infection is not so common.

Study populations of the lesser celandine to see if the leaves are infected and count the number of leaves infected in several samples of 100. Repeat this for a second or third area. Group the

Plate 8. (*top*) Lesser celandine (*Ranunculus ficaria*) and (*bottom*) wood sorrel (*Oxalis acetosella*).

Plate 9. (*left*) A birch tree infected by *Polyporus betulinus* and (*right*) the fruiting body of *P. betulinus*.

samples in a series increasing by 5% at each step and make a histogram of the number of samples in each group against the percentage infected.

The significance of this particular set of figures can be investigated by statistical methods. This investigation can be made the

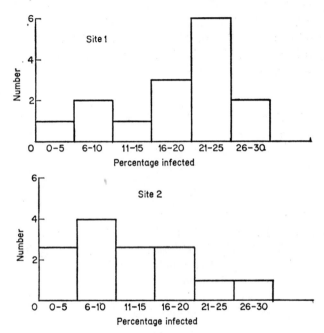

Figure 9. The infection of *Ranunculus ficaria* v. *ficaria* by the rust fungus *Uromyces ficariae*.

basis for further studies of this disease; also there are two well marked forms of the lesser celandine, one v. *fertilis* and the other v. *ficaria*; v. *fertilis* is the common plant, especially in well lit areas. The leaves do not have bulbils in their axils and the petals are broad and overlapping; the number of stamens is about 19–60, the number of carpels 11–72 and the plant sets fertile seed. v. *ficaria* occurs more frequently in the shade; when fully grown

D

the leaves develop small bulbils in the leaf axils which grow directly into new plants. The flower is a little smaller than that of v. *fertilis* and the petals are narrow and they do not normally overlap. The stamens number 14–40, the carpels 5–44, but the plants yield only 0–6 seeds per head. Incidentally v. *ficaria* is a tetraploid with chromosome number $2n = 32$ while v. *fertilis* has $2n = 16$. Investigations can be made to see if these two are equally susceptible to the disease. Alternatively the biology of the rust fungus is worthy of investigation. How does the fungus survive over winter, for example?

Investigation 26: To study the infection of birch trees by *Polyporus betulinus*

Birch trees are frequently infected by a fungus *Polyporus betulinus* (see Plate 9), the fructification of which first appears as a roundish white knob on a branch or the trunk. It rapidly expands to a bracket with a pale brown upper surface and a white underside consisting of many minute pores. The fungus causes a decay of the sap wood and very often the tree snaps off about 8–10 feet from the ground. Typically the *Polyporaceae* are wound parasites, the spores gaining entry only through the broken ends of branches.

To determine the frequency of infection in different habitats, it is only necessary to record the number of trees infected out of the total growing in any habitat, infection being judged by the presence of a fructification. It may well be that other trees are infected without producing a fructification but there is no easy means of recognising them. There are two species of birch (*Betula pendula* and *pubescens*) which are common in this country and comparisons may be made to see if there is any differential infection.

It is also of interest to investigate the age at which the fructification develops. This can be approximately achieved by measuring the diameter of all the trees in a woodland and recording the number infected. The diameter should be measured at eye-level and where the tree has more than one trunk springing from a

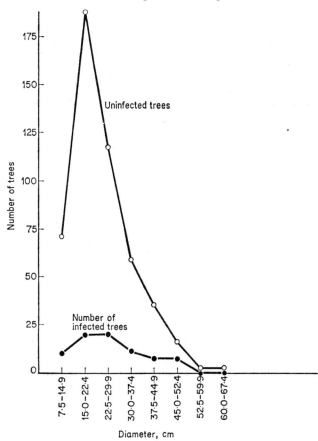

Figure 10. The infection of birch trees by *Polyporus betulinus*.

basal stool, each trunk should be treated as a separate individual.
The diameters should be grouped and the numbers of each plotted
as a curve. What conclusions can be drawn from the data shown
in Figure 10?

References

Crystal, R. N. (1948). *Insects of the British Woodlands.* Warne.

Smith, A. J. E. (1963). Variation in *Melampyrum pratense. Watsonia,* 5, 336–367.

Watt, A. S. (1919). On the causes of failure of natural regeneration in British Oakwoods. *J. Ecol.,* 7, 173–203.

Yeo, P. F. (1961). Germination, Seedlings and Formation of Haustoria in *Euphrasia. Watsonia,* 5, 11–22.

5

Dispersal and establishment in woodland plants

Many species of plant show remarkably efficient methods of dispersal. Attention is drawn to splendid examples like the plumed fruits of the *Compositae*, the winged seeds of the *Caryophyllaceae*, the hooked receptacles of agrimony and many others. There are, however, some very successful species that apparently lack an efficient means of dispersal. Acorns, hazel nuts, horse chestnuts merely fall from the tree and it is obvious that they are not carried about in the same way as ash, sycamore, lime or elm. Acorns and hazel nuts are, however, relatively large and contain a bulky food reserve which is, in part, a compensation for poor dispersal. Sir E. J. Salisbury (1942) has drawn attention to the fact that, on the whole, the larger seeds and fruits belong to plants of closed communities, whereas those with very light seeds tend to belong to open communities and to appear in the early stages of plant successions. Thus the oak invades communities already well established and needs a large food reserve to tide it over the early stages when there is much competition. There is therefore, an approximate relationship between the size of the seed and the type of community in which the plant grows.

Investigation 27: The dispersal and establishment of the oak

Search oak woodlands for the young (0–5 years old) oak seedlings; record the distances from the nearest oak tree. Very slight observation will show that a vast majority of the acorns fall below the tree and that dispersal over a considerable distance only occurs to a small fraction of the total, and that only a very small proportion of the whole ever survive to become small seedlings.

In order to determine the fate of the acorns the following experiments may be carried out:

(*a*) Mark out a small area of bare ground (1 square metre) below an oak tree and count the number of acorns beneath it. Since the leaves of the oak fall after the acorns many of the latter are covered and must be searched for. Record the numbers at weekly intervals through the autumn, when it will be found that they disappear.

(*b*) Mark out an area of bare ground in a woodland, place 100 sound acorns on the surface and 100 just below the surface. Cover the area with wire netting (small mesh) and record the number of acorns that disappear within a month.

(*c*) Repeat experiment (*b*) but bury the wire netting round the edge to a depth of 20 cm as well as covering the surface. Use netting fine enough to exclude burrowing animals such as mice and record your results.

(*d*) Previous experiment (Investigation 22) has shown the germination rate of acorns to be high, yet observation in woodlands shows a number of acorns which appear to be quite sound yet do not germinate while others germinate immediately on falling from the tree. Water content and the position of the micropyle in relation to the water supply are important.

Germinate 100 acorns placed vertically in the soil
Germinate 100 acorns placed the other way up
Germinate 100 acorns placed sideways in the soil

Germinate also 100 acorns under optimum conditions which have been dried in the laboratory so as to lose 10% of their fresh weight. Repeat with a second sample that have been dried to lose 20% and a third that have been dried so as to lose 30%.

From these experiments summarise clearly the factors that affect the germination of the acorns. Is there, for instance, any significance in the fact that the leaves of the oak fall after the acorns rather than the reverse?

Germination is but the first stage in the growth of the oak, and seedlings are liable to many hazards. Carry out similar experiments to (c) with seedlings to see what the result is of enclosing them in a cage of wire netting.

Investigation 28: To study dispersal in trees possessing winged fruits, e.g. sycamore and ash.

It is obvious that sycamore and ash possess more effective methods of dispersal than the oak. Try to gain some exact information by taking 100 fruits and painting them red, and then releasing them from a height (e.g. top of a school building) in various strengths of wind (measured with an anemometer). In addition, if an isolated tree can be found, paint a number of fruits *in situ*, and then by searching for them see how far they travel. The necessary observations will take longer in the case of a tree like the ash, where the fruits only fall from the tree over a fair period of time. Another way of comparing the two mechanisms is by directly measuring the rates of fall of the two kinds of fruit as a simple laboratory experiment. (Ashby, 1969.)

The structure of the fruits and enclosed seeds provides a useful morphological study and the germination of them can be studied in a similar manner to that of the acorn. Sycamore seed will not germinate before about mid-January; in nature radicles are produced in early February if the temperature is high. Therefore study the germination by sowing samples in February, March, April and May. Patience is required for the investigation of the germination of the ash; exposure to low temperature is necessary, and only about 5% germinates after the first winter, the majority only growing after exposure to two winters.

Investigation 29: To study dispersal in the herb bene't *(Geum urbanum)*

Geum urbanum is a plant of woodland paths, hedge-banks and similar places, occurring most frequently on the damper and better soils. It is dispersed by its hooked fruits, the hooks being formed from the kinked styles. Record the distribution of this species in part of a woodland, by showing individuals on large-scale maps. Generally, its distribution may be related to (*a*) effective dispersal along pathways and (*b*) the different environment of the pathway as compared with the remainder of the woodland. Investigate these possibilities by sowing seed away from the pathways in (*a*) areas cleared of other vegetation and (*b*) the normal field layers.

Investigation 30: To study the patterns of germination shown by woodland species

(*a*) *Bluebell*

Bluebell plants produce about 100 viable seeds per plant and the seeds lose their capacity for germination if allowed to dry completely. They must be stored in a practically saturated atmosphere, say, above a water surface in a control jar. Divide the sample into two; treat one half for 8–10 weeks in a refrigerator at about 4° C and keep the other as a control. Germinate both samples for a considerable period of time, observing and recording germination from week to week.

(*b*) *Dog's mercury*

Collect some seed and carry out germination tests; these will probably show a low rate of germination. Correlate this with the behaviour of the plant in nature and with the frequency of the occurrence of seedlings in the wild.

(*c*) *Lesser celandine*

At the time of separation from the parent, the seed of the lesser celandine contains only a small undifferentiated embryo. Development proceeds slowly through the autumn and winter and is complete only just before the spring. Compare the two sub species (p. 35) as regards (*a*) seed production and (*b*) germination. Germination in this species is independent of light and takes place in about 10–20 weeks.

(*d*) *The plants of recently coppiced areas, e.g. foxglove*

Germinate samples of 100 seeds of foxglove on damp filter paper in petri dishes placed (*a*) in the light and (*b*) in darkness. Compare the results in the two samples and repeat this with other species of plant, e.g. mullein (*Verbascum thapsus*), the marsh thistle (*Cirsium palustre*) which are known to grow well in woodland in the years immediately following coppicing.

Investigation 31: The effect of chemical substances on germination

It is well known that certain chemicals affect the germination of seeds. Thus coumarin, the substance responsible for the smell of newly mown hay, inhibits the germination of many seeds. Try, for example, the effect of a 1 % solution of this on the germination of lettuce, rye grass or cress. In each case, germinate 100 seeds as a control in a petri dish and compare carefully with a similar sample to which the coumarin has been added.

The pulp of berries contains substances that inhibit germination, and presumably this is one of the reasons why seeds do not germinate in the berry. Carry out similar experiments to those outlined above, but using the diluted juice of the woody nightshade (*Solanum dulcamara*).

Some seeds are known to produce substances that inhibit the growth of seeds near by. Rye grass (*Lolium perenne*) can suppress the germination of the common mayweed (*Tripleurospermum maritimum*), see Plate 7. Try this by sowing 100 seeds of each on petri dishes as controls and 100 of each together on the same dish. Roots, too, are known to produce substances which can suppress seed germination; the best known are the roots of grasses like couch (*Agropyron repens*) and red fescue (*Festuca rubra*). Grow the seeds of these grasses in petri dishes, and then sow among the roots, the seeds of other plants, e.g. cress and radish, taking care however that each experiment has adequate controls. Here, then, is a whole field of work for further investigation.

References

Ashby, M. (1969). *Introduction to Plant Ecology*. Macmillan.
Salisbury, E. J. (1942). *The Reproductive Capacity of Plants*. Bell.
Watt, A. S. (1919). On the causes of failure of natural regeneration in British Oakwoods. *J. Ecol.*, 7, 173–203.

Index